Who Am I?

Name the figure.

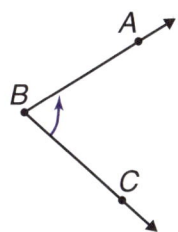

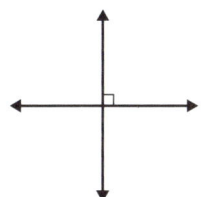

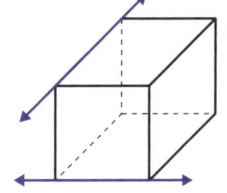

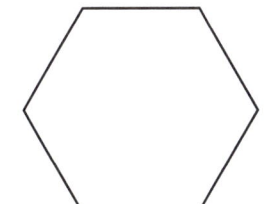

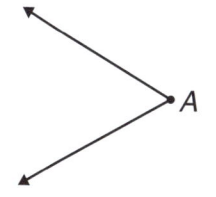

1. Circle
2. Angle
3. Line segment
4. Ray
5. Plane
6. Line
7. Point
8. Parallel lines
9. Perpendicular lines
10. Skew lines
11. Polygon
12. Vertex

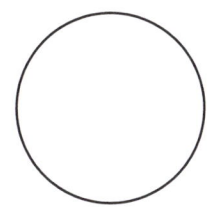

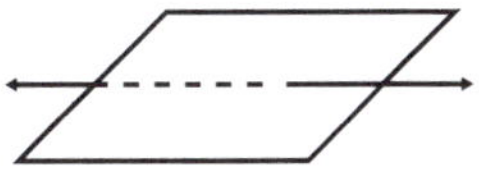

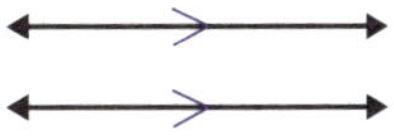

Answer Box

A	B	C	D	E	F
Line	Parallel lines	Plane	Polygon	Point	Circle
G	H	I	J	K	L
Vertex	Angle	Perpendicular lines	Line segment	Ray	Skew lines

Objective: Match a geometric figure to its name.

What's My Angle?

Measure the angle.

1.
2.
3.
4.
5.
6.
7.
8.
9.
10.
11.
12.

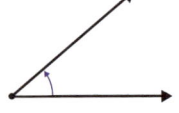

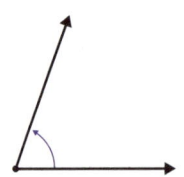

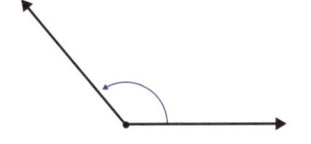

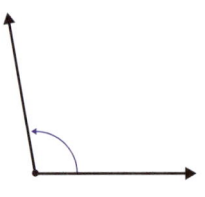

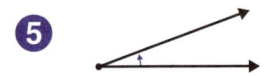

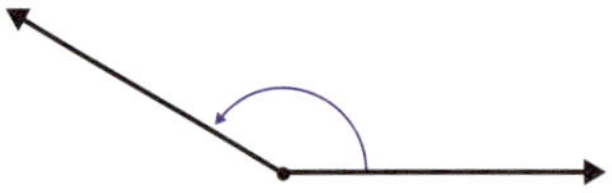

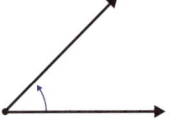

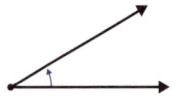

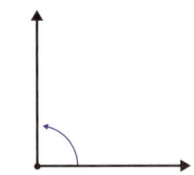

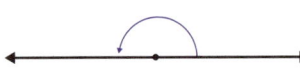

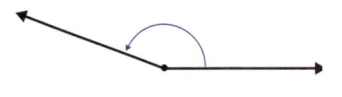

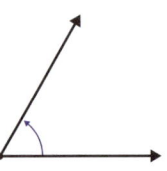

Use an angle ruler or a protractor.

Answer Box

A	B	C	D	E	F
100°	40°	130°	90°	150°	70°
G	H	I	J	K	L
180°	45°	160°	20°	30°	60°

2 Objective: Measure an angle.

Classified!

Look at the interior angles in the figure. Name the angle that matches the description.

1. acute
2. obtuse
3. right
4. straight

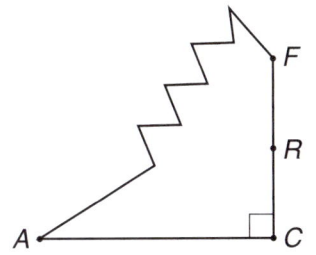

5. acute
6. straight
7. obtuse
8. right

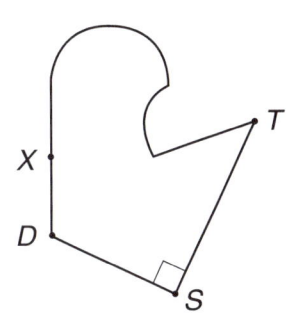

9. right
10. straight
11. obtuse
12. acute

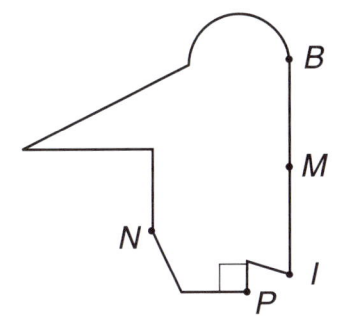

Answer Box

A ∠P	B ∠T	C ∠D	D ∠M	E ∠F	F ∠I
G ∠N	H ∠C	I ∠A	J ∠R	K ∠S	L ∠X

Objective: Classify an angle as acute, right, obtuse, or straight.

Matching Fun!

Find the most specific name of the geometric shape.

1. 2-dimensional closed figure with many straight sides
2. 4-sided figure
3. 4-sided figure with only one pair of parallel sides
4. figure with 4 right angles and 4 equal sides
5. figure with 8 sides
6. figure with 5 sides
7. 6-sided figure
8. 3-sided figure

9. 
10. 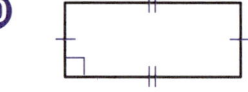
11. 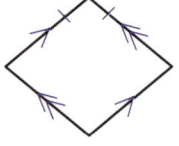
12.

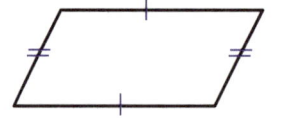

The prefixes of the words in the Answer Box can help you.

Answer Box

A	B	C	D	E	F
Triangle	Pentagon	Hexagon	Polygon	Rhombus	Trapezoid
G	H	I	J	K	L
Quadrilateral	Parallelogram	Square	Rectangle	Kite	Octagon

4 Objective: Identify a 2-dimensional figure, given its attributes.

All Kinds of Triangles

Choose the triangle that is described.

1. acute — X or S
2. obtuse — P or O
3. right — W or M
4. equilateral — N or T
5. scalene — Q or R
6. isosceles — O or U
7. isosceles right — M or V
8. isosceles acute — V or W
9. scalene right — X or S
10. scalene obtuse — N or P
11. acute — M or Q
12. obtuse — T or U

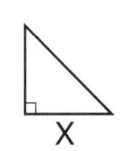

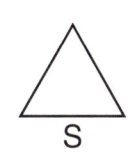

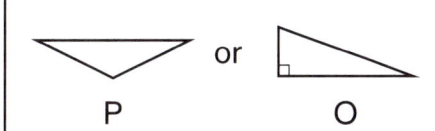

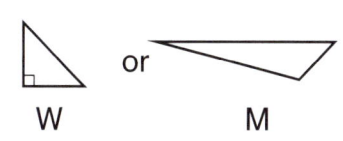

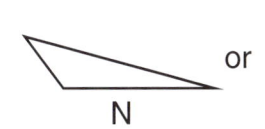

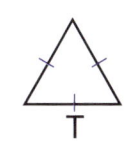

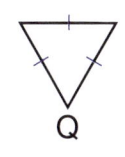

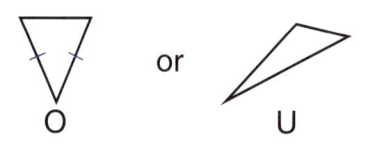

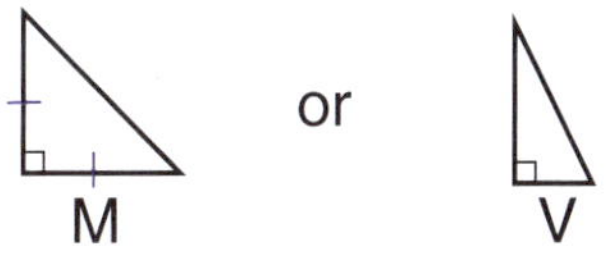

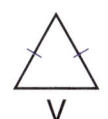

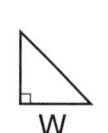

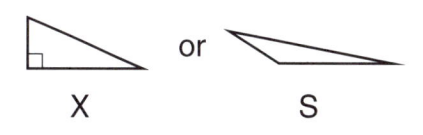

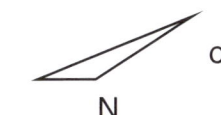

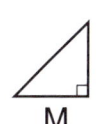

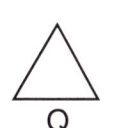

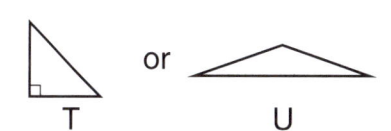

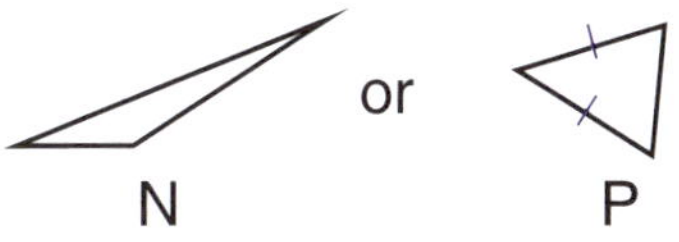

Answer Box

A	B	C	D	E	F
M	N	O	P	Q	R

G	H	I	J	K	L
S	T	U	V	W	X

Objective: Identify a triangle by its angles or sides.

Going in Circles!

Use the circle to identify the part or find the answer.

1. $\overline{RS}$
2. ∠SCD
3. $\overline{CD}$
4. point C
5. $\overline{RD}$

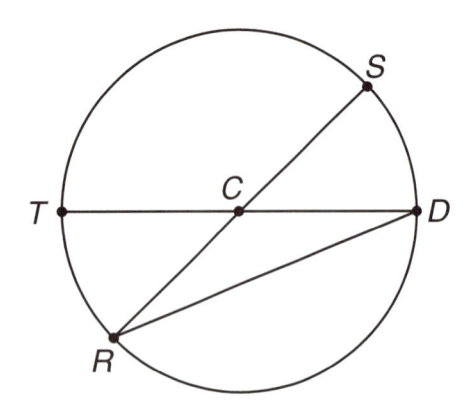

6. The measure of $\overline{TC}$ equals the measure of ■.

7. A radius of a circle can never be a chord. True or false?

8. The measure of $\overline{CD}$ is half the measure of ■.

9. If the measure of ∠RCT is 45°, then the measure of ■ also equals 45°.

10. If the measure of ∠RCD is 135°, then the measure of ■ also equals 135°.

11. A ■ is half of a circle.

12. A chord can never be a diameter of a circle. True or false?

Answer Box

A	B	C	D	E	F
Radius	∠DCS	$\overline{CS}$	Central angle	Center	Diameter

G	H	I	J	K	L
Chord	Semicircle	False	$\overline{TD}$	True	∠TCS

Objective: Identify and name the parts of a circle.

You Look Familiar!

Find the two congruent shapes.

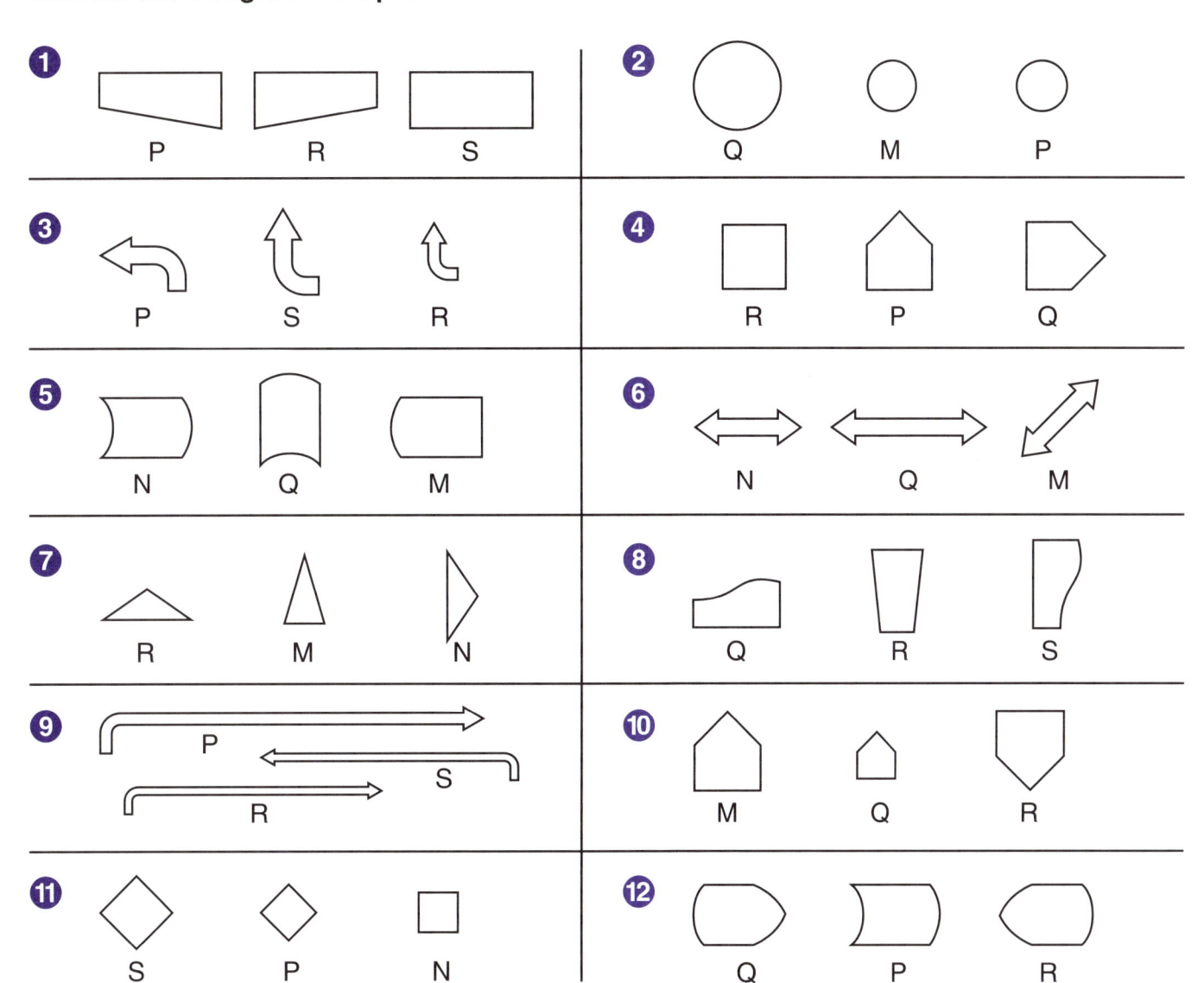

1. P, R, S
2. Q, M, P
3. P, S, R
4. R, P, Q
5. N, Q, M
6. N, Q, M
7. R, M, N
8. Q, R, S
9. P, S, R
10. M, Q, R
11. S, P, N
12. Q, P, R

Answer Box

A	B	C	D	E	F
M and N	M and P	N and P	N and Q	P and Q	P and R
G	H	I	J	K	L
Q and R	Q and S	R and S	P and S	N and R	M and R

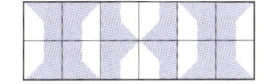

Objective: Match congruent figures.

7

Mirror Images!

Remember! A **reflection** is a flip, and a **translation** is a slide.

Choose the figure that shows a reflection.

1. 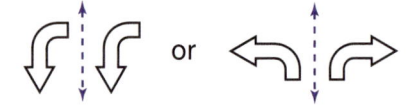
2. 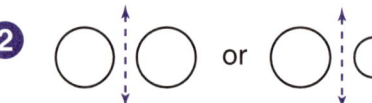
3. 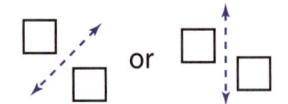
4. 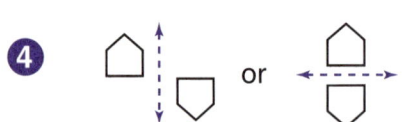
5. 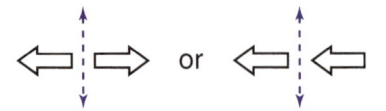
6.

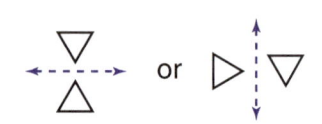

Choose the figure that shows a translation.

7. 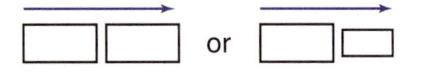
8. 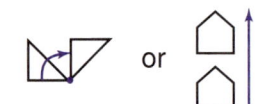
9. 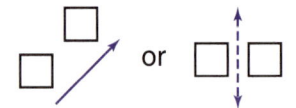
10. 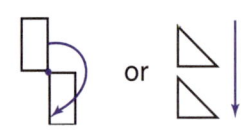
11. 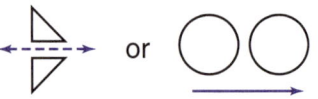
12.

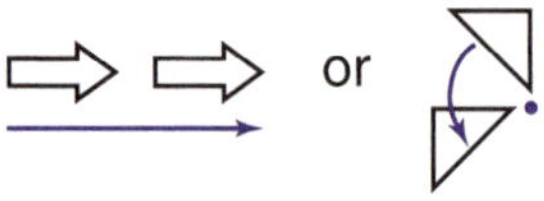

Answer Box

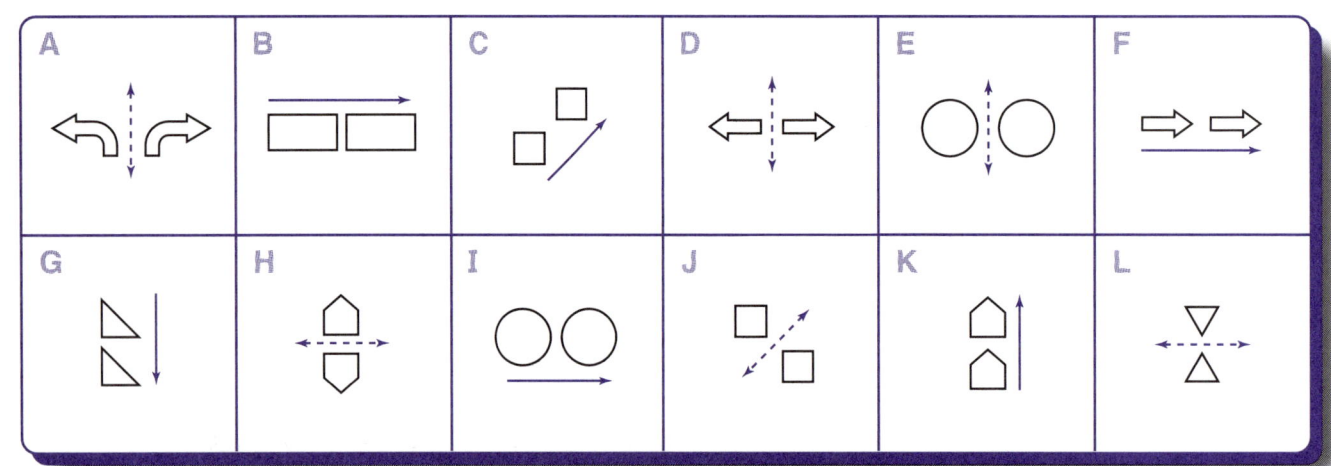

A	B	C	D	E	F
G	H	I	J	K	L

8 **Objective:** Identify a figure as a translation or a reflection of a given figure.

Round About!

Choose the figure that shows a rotation.

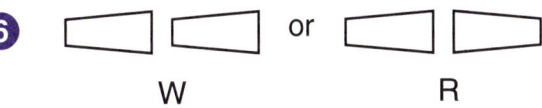

1. 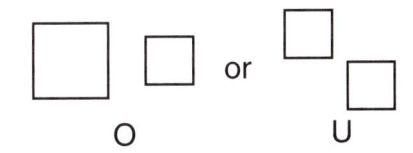 O or U
2. 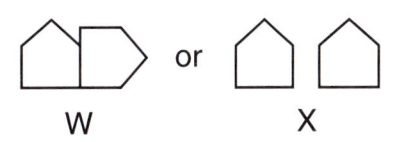 W or X
3. 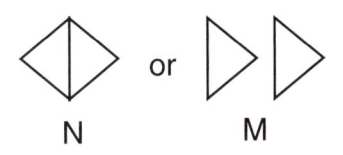 N or M
4. 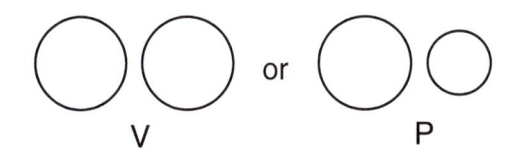 V or P
5. 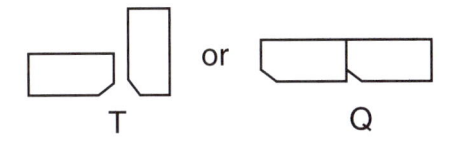 T or Q
6. W or R
7. 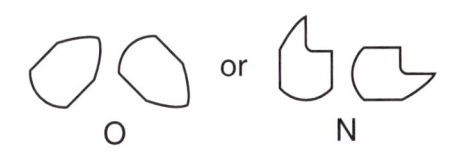 O or N
8. 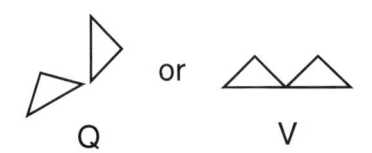 Q or V
9. 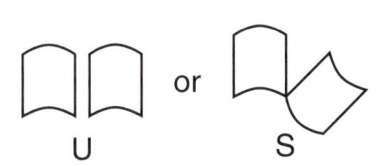 U or S
10. 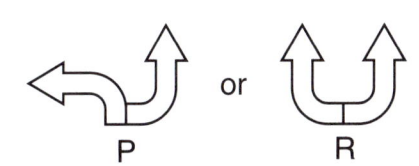 P or R
11. 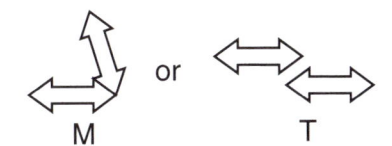 M or T
12. 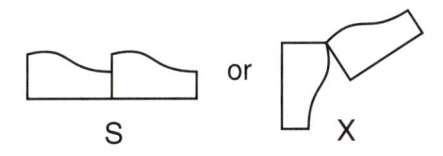 S or X

Answer Box

A	B	C	D	E	F
M	N	O	P	Q	R
G	H	I	J	K	L
S	T	U	V	W	X

Objective: Identify a figure that is a rotation of a given figure.

9

Problem Solving: Using a Drawing

Use a drawing to help you solve the problem.

1. Four students line up to buy tickets to the fair. Jean buys the first ticket. Jorge buys his ticket before Kenny. Kelly is third in line. Who is last in line?

2. Who is second in line in problem 1?

3. At the fair, a quilt that uses a pattern of circles and squares is displayed. Inside each square, there is a circle that touches each side of the square at one point. If the side of a square measures 8 in., then how many inches is the radius of the circle?

4. Another quilt pattern uses pentagons and squares. To make the design, the quilter uses one pentagon and then puts a square on each side of the pentagon. How many sides does the new shape have?

5. The new shape in problem 4 has a hexagon in it. True or false?

6. A square box holds 4 jars of homemade jam. Each jar fits tightly against 2 sides of the box and 2 other jars. The box measures 10 in. on each side. About how many inches is the diameter of a jar of jam?

7. A display about honey shows the hexagon shape of a honeycomb. What is the total number of diagonals in the hexagon?

8 The total number of diagonals in a pentagon is 5. True or false?

9 The craft exhibits are displayed on long tables in rows 26 ft long. Each table is 8 ft long. How many feet are there between each table along the row if the tables are evenly spaced and there are 3 tables in each row?

10 How many rows are needed to set up 20 tables in problem 9?

11 There are 3 goats being judged. In how many different ways can they be lined up in a row?

12 Small animal pens are set up along an 18-ft-long walkway inside the livestock area. Each square pen measures 4 ft on a side, and is 3 ft away from the next pen. How many pens can there be on one side of the walkway?

Answer Box

A	B	C	D	E	F
True	3	Kenny	15	6	False
G	H	I	J	K	L
Jorge	5	7	9	1	4

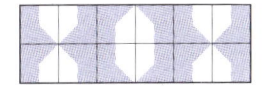

Objective: Solve a problem by using a drawing.

Measure Up!

Measure the line segment in inches.

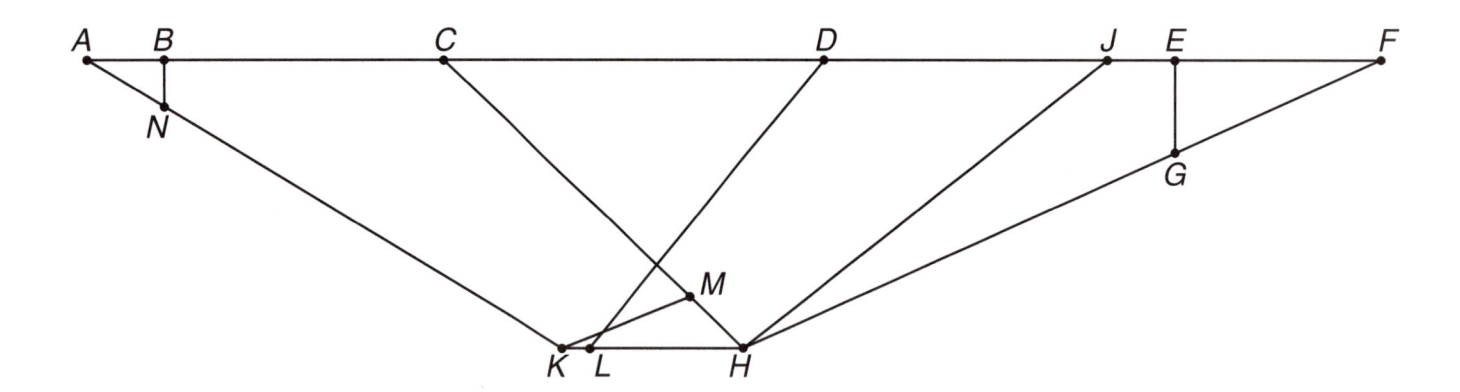

1. $\overline{AF}$
2. $\overline{AD}$
3. $\overline{BN}$
4. $\overline{AK}$
5. $\overline{LD}$
6. $\overline{KH}$
7. $\overline{HF}$
8. $\overline{GE}$
9. $\overline{HJ}$
10. $\overline{CH}$
11. $\overline{JF}$
12. $\overline{KM}$

Many of the line segments have fractional measurements.

Answer Box

A	B	C	D	E	F
$3\frac{3}{4}$ in.	$2\frac{1}{4}$ in.	$\frac{1}{2}$ in.	$\frac{3}{4}$ in.	2 in.	$1\frac{1}{2}$ in.
G	H	I	J	K	L
7 in.	3 in.	4 in.	1 in.	$2\frac{1}{2}$ in.	$\frac{1}{4}$ in.

Objective: Measure a line segment, using a customary ruler.

That Makes Sense!

Choose the best estimate.

1. 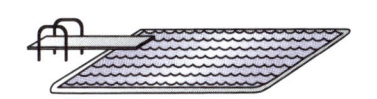 5 qt or 50,000 gal

2.  5 pt or 5,000 gal

3. 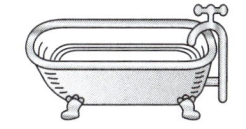 50 gal or 5 fl oz

4. 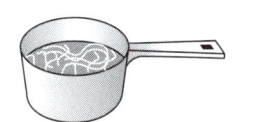 5 qt or 50,000 gal

5.  50 gal or 5 pt

6.  5 pt or 5 fl oz

7.  5 lb or 5 oz

8.  8 lb or 8 oz

9.  250 lb or 2 T

10. 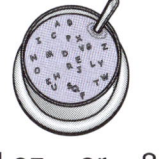 8 fl oz or 8 lb

11.  250 lb or 2 T

12.  5 oz or 5 lb

Answer Box

A	B	C	D	E	F
50 gal	250 lb	5 fl oz	5,000 gal	5 lb	50,000 gal
G	H	I	J	K	L
5 pt	2 T	8 lb	5 oz	5 qt	8 fl oz

Objective: Determine the reasonable customary unit of capacity or weight for a given container or object.

13

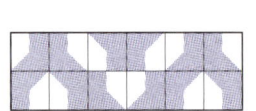

Give It Another Name!

Complete the statement.

1. 6 yd = 216 ▇
2. 3 mi = ▇ ft
3. 7 c = ▇ fl oz
4. ▇ gal = 24 pt
5. ▇ T = 16,000 lb
6. ▇ oz = 3 lb
7. 144 in. = 4 ▇
8. 1,760 yd = 5,280 ▇
9. 5 ▇ = 80 fl oz
10. 36 ▇ = 144 c
11. ▇ pt = 3 gal
12. ▇ oz = 6 lb

Answer Box

A	B	C	D	E	F
ft	96	pt	56	qt	yd
G	**H**	**I**	**J**	**K**	**L**
48	15,840	8	24	3	in.

Complete the statement.

1. ▇ c = 6 qt
2. 32 oz = ▇ lb
3. 72 in. = 6 ▇
4. 5,280 yd = ▇ mi
5. 9 yd = ▇ ft
6. ▇ T = 32,000 lb
7. ▇ gal = 32 pt
8. 24 qt = 6 ▇
9. 6 c = ▇ fl oz
10. 16 ▇ = 4 gal
11. 24 ▇ = 72 ft
12. 128 oz = ▇ lb

Answer Box

A	B	C	D	E	F
24	3	8	gal	ft	48
G	**H**	**I**	**J**	**K**	**L**
4	qt	2	16	yd	27

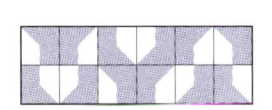

Objective: Convert between customary units of length, capacity, and weight.

Make Mine Metric!

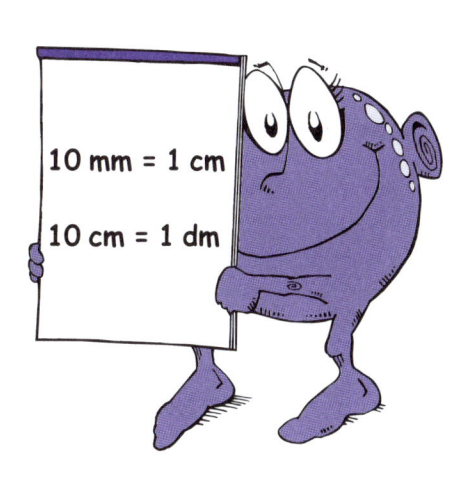

Measure the line segment.

1. ET = ▩ mm
2. RS = ▩ cm
3. EB = ▩ dm
4. RT = ▩ mm

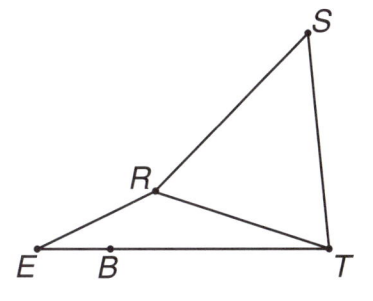

10 mm = 1 cm
10 cm = 1 dm

5. AB = ▩ dm
6. AK = ▩ cm
7. KY = ▩ cm
8. PB = ▩ mm

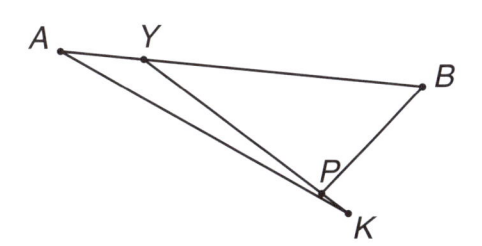

9. MN = ▩ mm
10. ML = ▩ cm
11. NT = ▩ mm
12. BT = ▩ cm

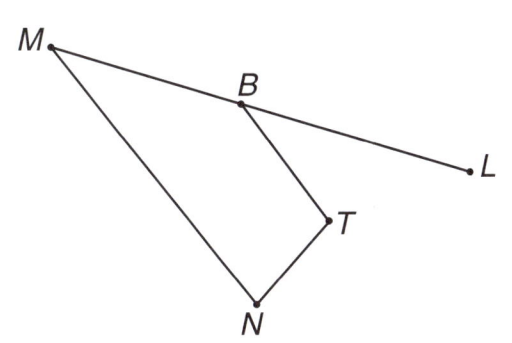

Answer Box

A	B	C	D	E	F
3.5	6	15	0.5	2	20
G	H	I	J	K	L
25	40	45	0.1	4.5	3

Objective: Measure a line segment, using a metric ruler.

15

Filled to Capacity!

Choose the best estimate.

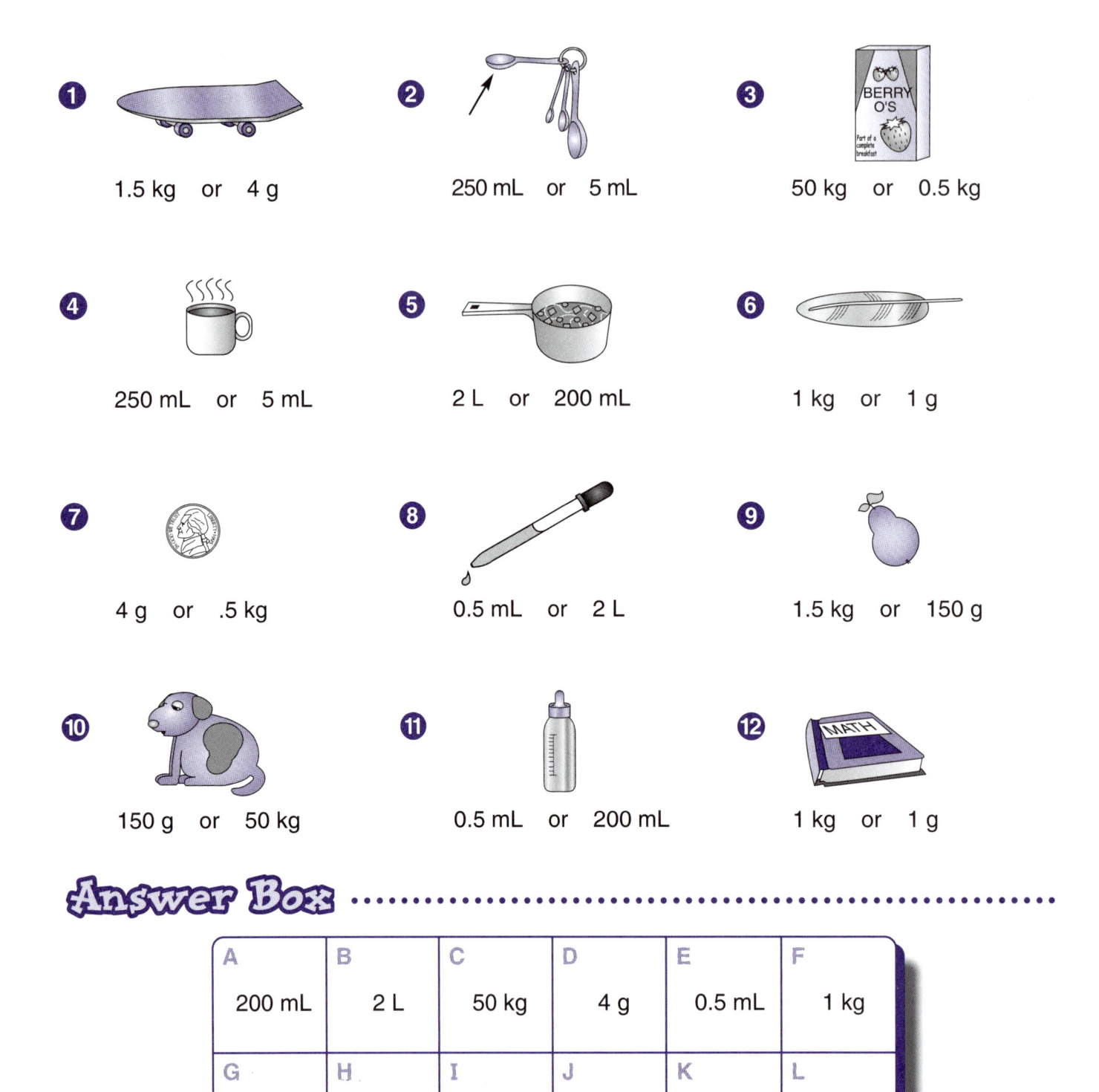

1. 1.5 kg or 4 g

2. 250 mL or 5 mL

3. 50 kg or 0.5 kg

4. 250 mL or 5 mL

5. 2 L or 200 mL

6. 1 kg or 1 g

7. 4 g or .5 kg

8. 0.5 mL or 2 L

9. 1.5 kg or 150 g

10. 150 g or 50 kg

11. 0.5 mL or 200 mL

12. 1 kg or 1 g

Answer Box

A	B	C	D	E	F
200 mL	2 L	50 kg	4 g	0.5 mL	1 kg
G	H	I	J	K	L
0.5 kg	150 g	250 mL	1.5 kg	1 g	5 mL

Objective: Determine the reasonable metric unit of capacity or mass for a given container or object.

16

More Names!

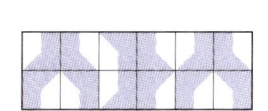

Find the answer.

1. ■ mm = 50 cm
2. 2.5 m = ■ cm
3. ■ mg = 2 g
4. ■ mL = 3 L
5. 4,500 mL = ■ L
6. 20 dm = ■ cm
7. ■ m = 5,000 cm
8. 2.4 kg = ■ g
9. 5 g = ■ mg
10. 4,000 mg = ■ g
11. 500 cm = ■ m
12. 25 mg = ■ g

Answer Box

A	B	C	D	E	F
2,400	0.025	500	50	5	4.5
G	H	I	J	K	L
250	200	4	3,000	5,000	2,000

Find the answer.

1. 42.3 mm = ■ cm
2. 396 g = ■ kg
3. 0.22 L = ■ mL
4. 0.016 cm = ■ dm
5. 3.5 km = ■ m
6. 12,632 g = ■ kg
7. 39.6 m = ■ dm
8. 1,600 mL = ■ L
9. 42.3 m = ■ cm
10. 22 kg = ■ g
11. 0.035 kg = ■ mg
12. 12,632 mm = ■ km

Answer Box

A	B	C	D	E	F
35,000	3,500	12.632	1.6	22,000	396
G	H	I	J	K	L
220	4,230	0.012632	0.396	0.0016	4.23

Objective: Convert between metric units of length, capacity, and mass.

Problem Solving: Solving Multi-Step Problems

Solve the problem.

1. A square vegetable and flower garden measures 25 ft on one side. A roll of fencing is 100 yd long. How many feet will be left on the roll after the perimeter of the garden is fenced?

2. Bags of soil are added to the vegetable beds. Each 50-lb bag usually costs $5.95. The garden center has a special price. If you buy 10 bags, then you get another bag free. About how much does a person save per bag when buying 11 bags at the special price?

3. Carrot seeds are very small, and 100 seeds weigh less than 1 oz. If a gardener buys two 2-oz packages, there will be at least 400 seeds to plant. True or false?

4. Beans have large seeds. A package can be planted in a row 18 ft long. The rows in the garden are 30 in. long. How many complete rows can be planted with one package of seeds?

5. Two gardeners began working in the garden at 1:45 P.M. and stopped at 4 P.M. How many hours did they work in the garden?

6. One harvest included 16 cucumbers, 34 tomatoes, and 5 lb of beans. The cucumbers each weighed about 10 oz and the tomatoes each weighed about 8 oz. What was the weight of the harvest to the nearest pound?

7. Some of the tomatoes are made into sauce. It takes 6 tomatoes to make 8 fl oz of sauce, which is enough to serve 2 people. How many tomatoes are needed to make enough sauce for 10 servings?

8 Some of the plants were eaten by bugs. In the garden, $\frac{1}{3}$ of the squash plants were destroyed by bugs. The garden had 12 plants, and the remaining plants produced 34 lb of squash. How many pounds of squash were probably lost because of the bugs?

9 One week the gardeners picked 5 bushels of strawberries. There are 4 pecks in a bushel and 8 qt in a peck. How many quarts of strawberries did they pick?

10 A nut tree in the yard produces 5 lb of nuts every fall. Only 20 oz of this harvest of nuts can be eaten. This accounts for 20% of the total number of nuts produced by the tree. True or false?

11 A gardener sets up a vegetable stand to sell some of what she has raised. She charges $1.40 per pound for beans and $1.25 for a dozen eggs. A customer orders 20 oz of beans and 2 dozen eggs. How much are the beans?

12 On one hot day in August, a gardener drank 2.5 bottles of water while working in the garden. How many cups of water is this if each bottle holds 8 fl oz?

Answer Box

A	B	C	D	E	F
$2\frac{1}{4}$	160	32	$0.54	30	200
G	H	I	J	K	L
True	$1.75	$2\frac{1}{2}$	17	7	False

Objective: Solve a multi-step problem involving units of measure.

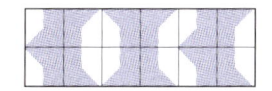

All the Way Around!

The **perimeter** (P) of any polygon equals the sum of the measures of the sides.

Find the perimeter of the figure.

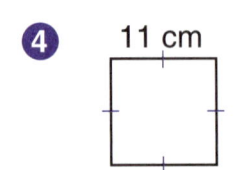

1. 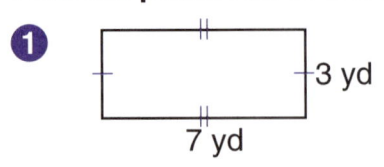 Rectangle: 3 yd, 7 yd
2. 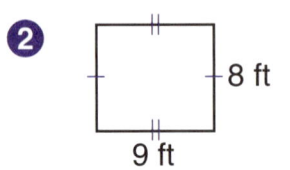 Rectangle: 8 ft, 9 ft
3. 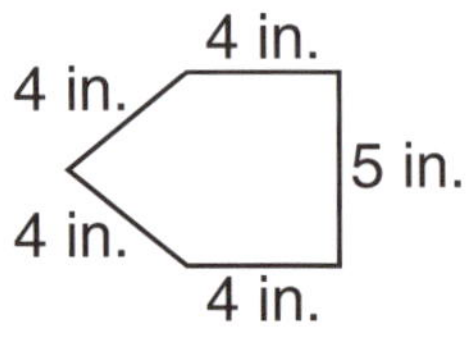 Pentagon: 4 in., 4 in., 5 in., 4 in., 4 in.
4. Square: 11 cm
5. 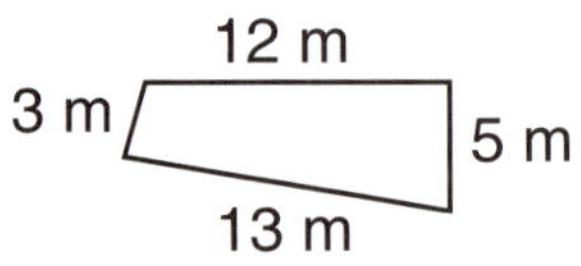 Quadrilateral: 3 m, 12 m, 5 m, 13 m
6. 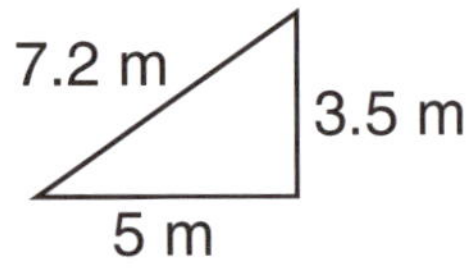 Triangle: 7.2 m, 3.5 m, 5 m
7. 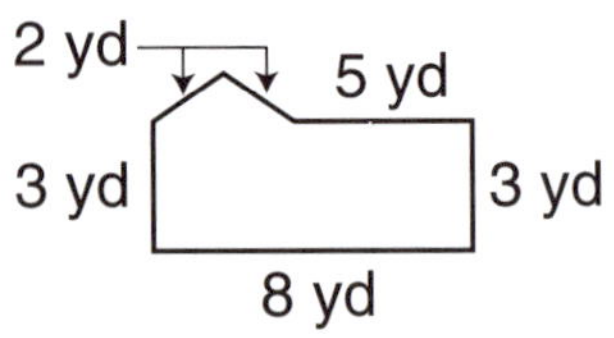 Figure: 2 yd, 5 yd, 3 yd, 3 yd, 8 yd
8. 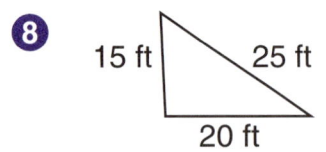 Triangle: 15 ft, 25 ft, 20 ft
9. 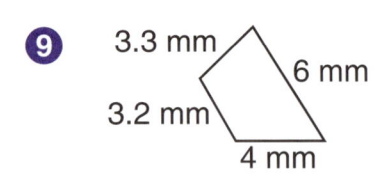 Quadrilateral: 3.3 mm, 6 mm, 4 mm, 3.2 mm
10. 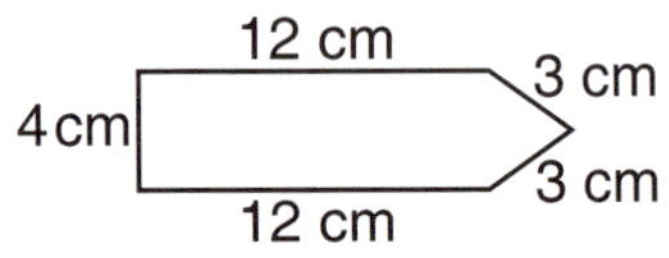 Figure: 12 cm, 3 cm, 3 cm, 12 cm, 4 cm
11. 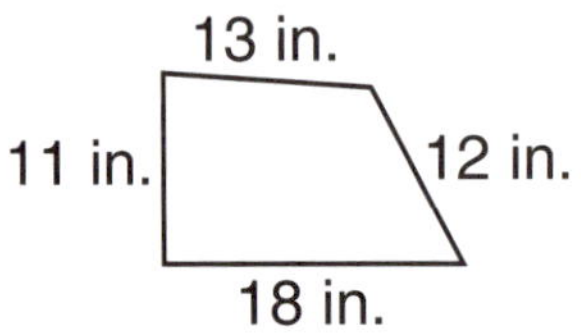 Quadrilateral: 13 in., 12 in., 18 in., 11 in.
12. Square: 4 mm

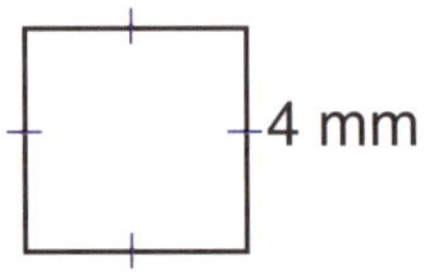

Answer Box

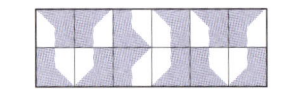

A	B	C	D	E	F
34 ft	15.7 m	54 in.	21 in.	20 yd	44 cm

G	H	I	J	K	L
16 mm	60 ft	16.5 mm	33 m	23 yd	34 cm

Objective: Find the perimeter of a given figure.

Going the Distance!

Find the perimeter of the figure.

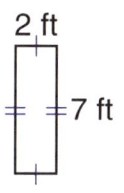

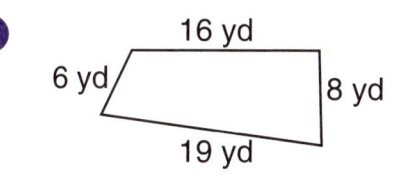

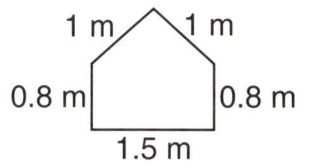

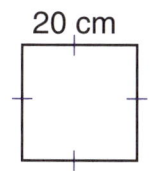

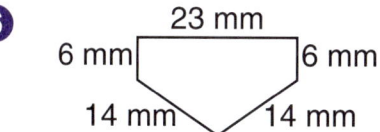

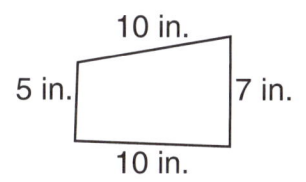

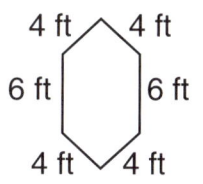

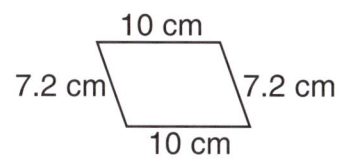

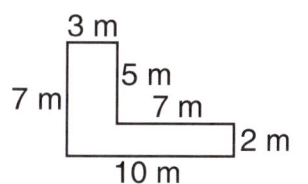

1. Rectangle: 6 in. by 8 in.
2. Rectangle: 2 ft by 7 ft.
3. Quadrilateral: 16 yd, 8 yd, 19 yd, 6 yd.
4. Pentagon: 1 m, 1 m, 0.8 m, 0.8 m, 1.5 m.
5. Square: 20 cm.
6. Pentagon: 23 mm, 6 mm, 6 mm, 14 mm, 14 mm.
7. Quadrilateral: 10 in., 7 in., 10 in., 5 in.
8. Hexagon: 4 ft, 4 ft, 6 ft, 6 ft, 4 ft, 4 ft.
9. Triangle: 14 yd, 4 yd, 17 yd.
10. Parallelogram: 10 cm, 7.2 cm, 10 cm, 7.2 cm.
11. L-shape: 3 m, 5 m, 7 m, 2 m, 10 m, 7 m.
12. Crown shape: 1 mm, 1.2 mm, 1 mm, 1 mm, 3 mm, 1 mm.

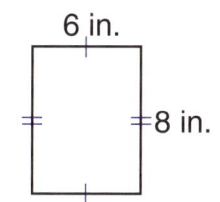

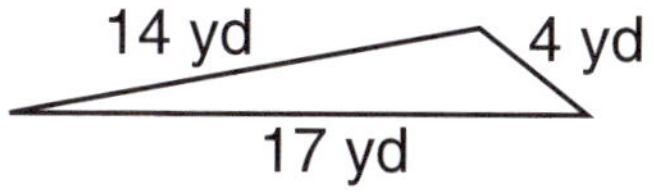

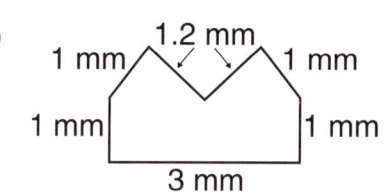

Answer Box

A	B	C	D	E	F
28 in.	32 in.	80 cm	34 m	28 ft	63 mm

G	H	I	J	K	L
34.4 cm	5.1 m	49 yd	35 yd	18 ft	9.4 mm

Objective: Find the perimeter of a given figure.

21

What's My Area?

CONSIDER THIS

To find the **area (A)** of a rectangle, multiply the **length (ℓ)** times the **width (w)**.

$A = \ell \times w$

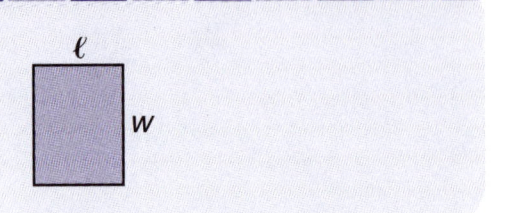

Find the area of the rectangle.

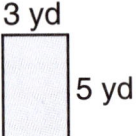

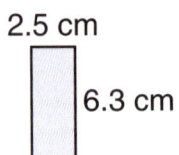

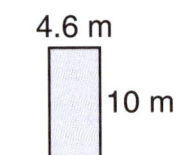

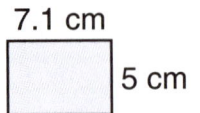

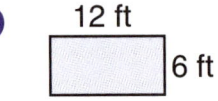

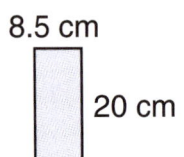

1. 3 yd, 5 yd
2. 2.5 cm, 6.3 cm
3. 4.6 m, 10 m
4. 7.1 cm, 5 cm
5. 12 ft, 6 ft
6. 5 in., 21 in.
7. 20 m, 8.5 m
8. 25 ft, 14 ft
9. 100 in., 75 in.
10. 120 yd, 75 yd
11. 8.5 cm, 20 cm
12. 25 m, 6.5 m

Answer Box

A	B	C	D	E	F
7,500 in.^{2}	46 m^2	15 yd^2	35.5 cm^2	15.75 cm^2	105 in.^{2}

G	H	I	J	K	L
72 ft^2	170 cm^2	170 m^2	9,000 yd^2	162.5 m^2	350 ft^2

Objective: Find the area of a rectangle.

Missing Measures

Find the missing measure of the rectangle.

1 ℓ = 7ft, w = ■, A = 63 ft²

2 ℓ = 5 in, w = ■, A = 125 in²

3 ℓ = ■, w = 6.9 m, A = 33.12 m²

4 ℓ = 23.5 cm, w = ■, A = 282 cm²

5 ℓ = ■, w = 12 yd, A = 96 yd²

6 ℓ = ■, w = 5.6 mm, A = 14 mm²

7 area = 84 m²
length = 7 m
width = ■

8 area = 300 in.²
length = ■
width = 30 in.

9 area = 125 yd²
length = ■
width = 10 yd

10 area = 14.7 cm²
length = ■
width = 3.5 cm

11 area = 160 mm²
length = 40 mm
width = ■

12 area = 78 ft²
length = ■
width = 3 ft

Area = length × width

Answer Box

A	B	C	D	E	F
2.5 mm	10 in.	9 ft	12 m	4.8 m	4 mm
G	H	I	J	K	L
26 ft	25 in.	4.2 cm	12 cm	8 yd	12.5 yd

Objective: Given the area, find the length or the width of a rectangle.

23

Triangles Ahead!

CONSIDER THIS

To find the **area (A)** of a triangle, multiply the **height (h)** by the **base (b)** and then divide by 2.

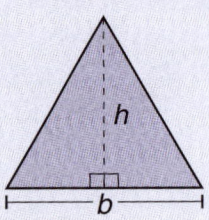

$$A = \frac{b \times h}{2}$$

Find the area of the triangle.

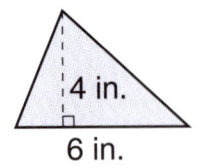

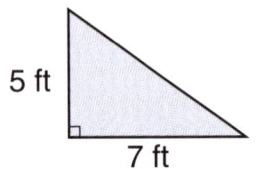

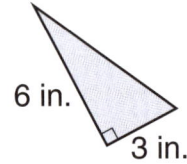

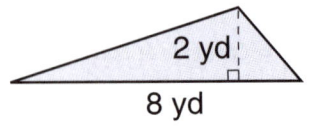

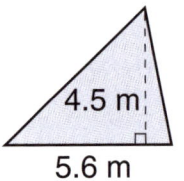

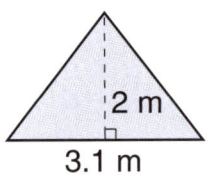

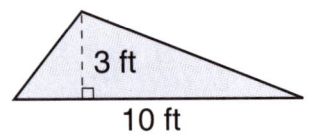

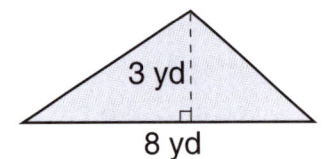

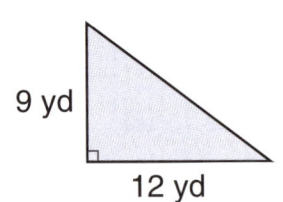

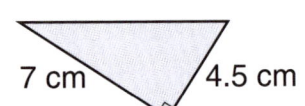

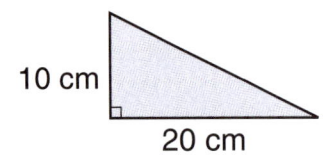

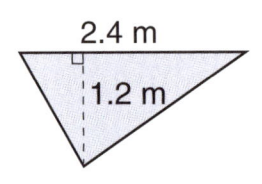

1. 4 in., 6 in.
2. 5 ft, 7 ft
3. 6 in., 3 in.
4. 2 yd, 8 yd
5. 4.5 m, 5.6 m
6. 2 m, 3.1 m
7. 3 ft, 10 ft
8. 3 yd, 8 yd
9. 9 yd, 12 yd
10. 7 cm, 4.5 cm
11. 10 cm, 20 cm
12. 2.4 m, 1.2 m

Answer Box

A	B	C	D	E	F
12 in.²	8 yd²	17.5 ft²	3.1 m²	54 yd²	9 in.²
G	H	I	J	K	L
15.75 cm²	15 ft²	1.44 m²	12 yd²	100 cm²	12.6 m²

24   Objective: Find the area of a triangle.

Fancy That!

> Remember! A **parallelogram** has two pairs of parallel and congruent sides.

CONSIDER THIS

To find the **area (A)** of a parallelogram, multiply the height by the base.

$A = b \times h$

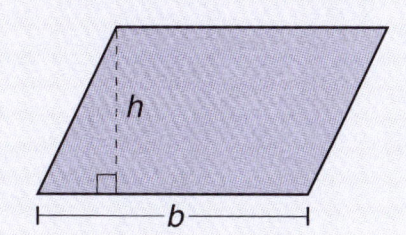

Find the area of the parallelogram.

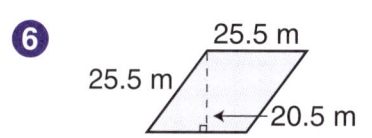

1. 12 in., 5 in., 3 in
2. 12 yd, 7 yd, 5 yd
3. 16 mm, 5 mm, 4 mm
4. 7 m, 9.9 m, 7 m
5. 1.8 mm, 2.7 mm, 2.5 mm
6. 25.5 m, 25.5 m, 20.5 m
7. base = 4.7 cm
   height = 5.2 cm
8. base = 25 in.
   height = 20 in.
9. base = 5 ft
   height = 48 ft
10. base = 10 cm
    height = 12 cm
11. base = 42 ft
    height = 6 ft
12. base = 36 yd
    height = 25 yd

Answer Box

A	B	C	D	E	F
252 ft²	64 mm²	500 in.²	522.75 m²	120 cm²	49 m²

G	H	I	J	K	L
240 ft²	4.5 mm²	60 yd²	900 yd²	36 in.²	24.44 cm²

Objective: Find the area of a parallelogram.

C Is for Circumference!

The radius is half the length of the diameter.

CONSIDER THIS

The **circumference (C)** of a circle equals π × d.

d represents the diameter of the circle.

C = π × d   π is approximately equal to (≈) 3.14

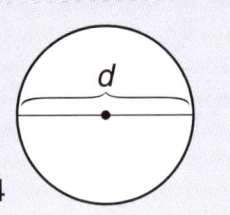

Find the circumference of the circle. Use π ≈ 3.14.

1. 6 ft
2. 3.1 cm
3. 5 in.
4. 50 ft
5. 6.5 in.
6. 12 mm
7. 25.5 m
8. 9 yd
9. 0.6 cm
10. 7.8 mm
11. 5 m
12. 10 yd

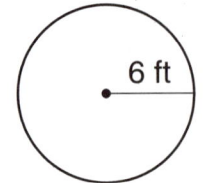

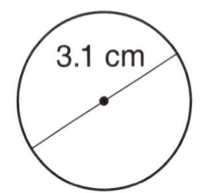

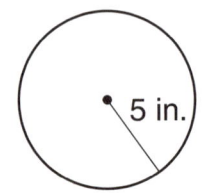

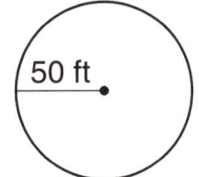

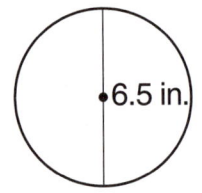

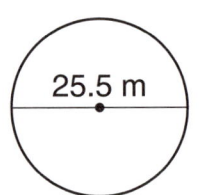

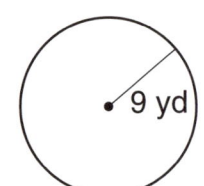

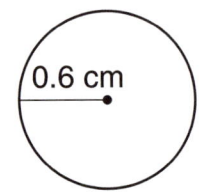

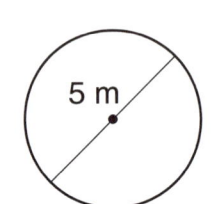

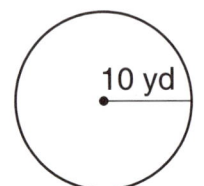

Answer Box

A	B	C	D	E	F
37.68 mm	9.734 cm	15.7 m	20.41 in.	37.68 ft	314 ft

G	H	I	J	K	L
56.52 yd	62.8 yd	3.768 cm	31.4 in.	80.07 mm	24.492 m

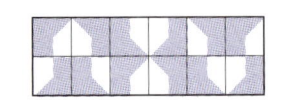

26   Objective: Find the circumference of a circle.

What's Inside?

CONSIDER THIS

The **area (A)** of a circle equals $\pi \times r^2$.

r represents the radius of the circle.

$A = \pi \times r^2$

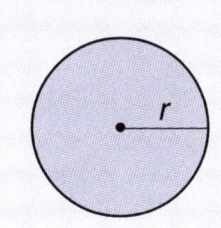

Find the area of the circle. Use $\pi \approx 3.14$.

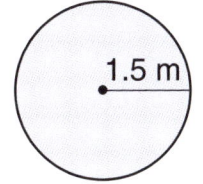

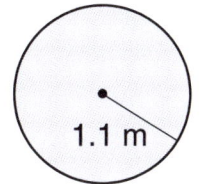

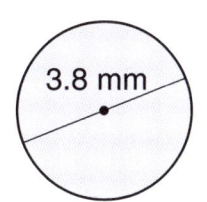

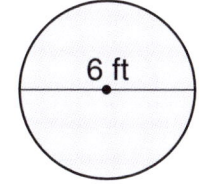

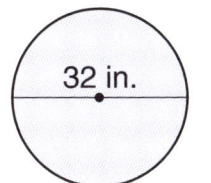

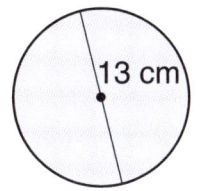

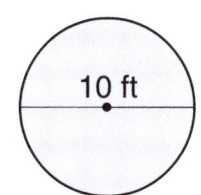

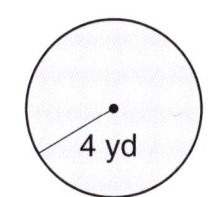

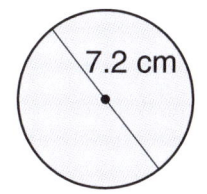

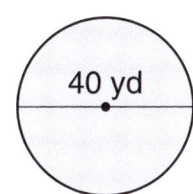

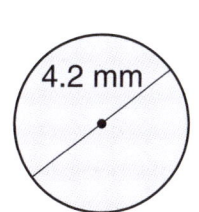

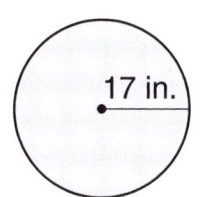

1. 1.5 m
2. 1.1 m
3. 3.8 mm
4. 6 ft
5. 32 in.
6. 13 cm
7. 10 ft
8. 4 yd
9. 7.2 cm
10. 40 yd
11. 4.2 mm
12. 17 in.

Answer Box

A	B	C	D	E	F
28.26 ft²	7.065 m²	803.84 in.²	13.8474 mm²	132.665 cm²	3.7994 m²

G	H	I	J	K	L
78.5 ft²	1,256 yd²	40.6944 cm²	11.3354 mm²	907.46 in.²	50.24 yd²

Objective: Find the area of a circle.

3-D Name Game!

Match the figure to its name or description.

1. 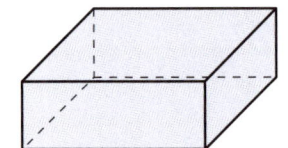
2. 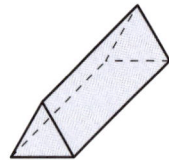
3. 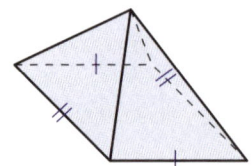
4. 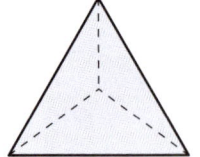
5. 
6.

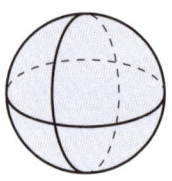

7. A ■ has 6 congruent faces.

8. A soup can is shaped like a ■.

9. A ■ has 5 faces, one of which is a square.

10. A ■ always has 6 faces shaped like a triangle.

11. A ■ is a sphere cut in half.

12. A ■ has 2 faces shaped like a hexagon and 6 faces shaped like a rectangle.

Answer Box

A	B	C	D	E	F
Hemisphere	Rectangular pyramid	Hexagonal pyramid	Cube	Triangular prism	Hexagonal prism
G	H	I	J	K	L
Square pyramid	Cone	Triangular pyramid	Rectangular prism	Cylinder	Sphere

28   **Objective:** Identify a 3-dimensional figure, given its attributes.

V Is for Volume!

Find the volume.

1. 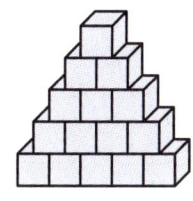
2. 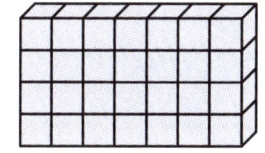
3. 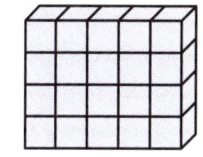
4. 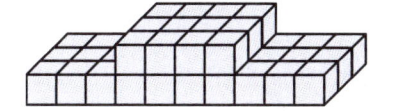
5. 
6. 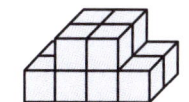
7. 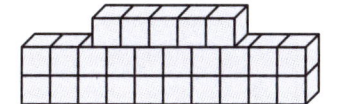
8. 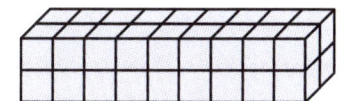
9. 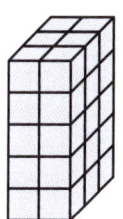
10. 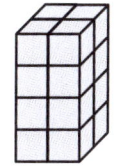
11. 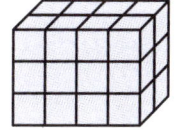
12.

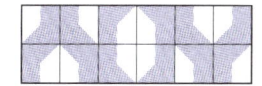

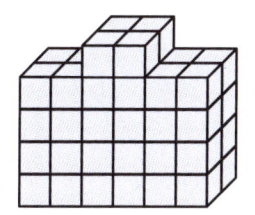

Answer Box

A	B	C	D	E	F
36 cubic units	52 cubic units	15 cubic units	42 cubic units	17 cubic units	24 cubic units

G	H	I	J	K	L
12 cubic units	28 cubic units	16 cubic units	25 cubic units	30 cubic units	20 cubic units

Objective: Find volume.

Turn Up the Volume!

> Remember! Volume is always measured in cubic units.

CONSIDER THIS

To find the **volume** of a rectangular prism or a cube, use this formula.

$V = \ell \times w \times h$

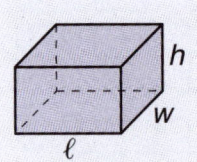

Find the volume of the rectangular prism.

1 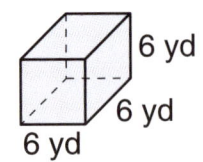 6 yd, 6 yd, 6 yd

2 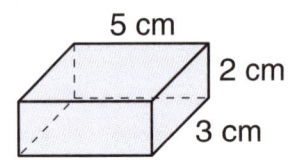 5 cm, 2 cm, 3 cm

3 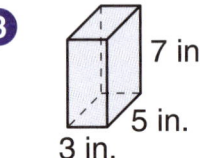 7 in., 5 in., 3 in.

4 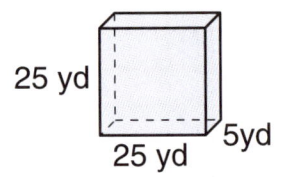 25 yd, 25 yd, 5 yd

5 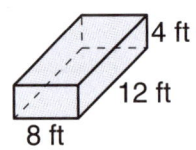 4 ft, 12 ft, 8 ft

6 10 m, 15 m, 27 m

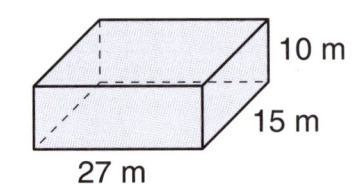

7 $\ell = 12$ mm
$w = 2$ mm
$h = 1$ mm

8 $\ell = 3$ cm
$w = 5$ cm
$h = 7$ cm

9 $\ell = 2$ mm
$w = 6$ mm
$h = 10$ mm

10 $\ell = 2$ ft
$w = 22$ ft
$h = 1$ ft

11 $\ell = 6$ m
$w = 23$ m
$h = 5$ m

12 $\ell = 10$ in.
$w = 10$ in.
$h = 10$ in.

Answer Box

A	B	C	D	E	F
24 mm³	44 ft³	4,050 m³	30 cm³	384 ft³	690 m³

G	H	I	J	K	L
3,125 yd³	216 yd³	105 cm³	1,000 in.³	105 in.³	120 mm³

30   Objective: Find the volume of a cube or a rectangular prism.

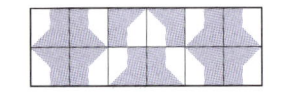

Draw It!

Find the actual distance represented by the lines, using the scale given.

Scale: 1 in. = 4 mi

1.
2.
3.
4.
5.
6.

Scale: 2 cm = 5 km

7.
8.
9.
10.
11.
12.

Answer Box

A	B	C	D	E	F
15 km	25 km	7 mi	1.25 km	12.5 km	2.5 km
G	H	I	J	K	L
2 mi	6 mi	12 mi	7.5 km	9 mi	4 mi

Objective: Determine the actual distance given a scale drawing.

Saving Space with Scale

Find the scale drawing length for the given actual distance, using the scale given.

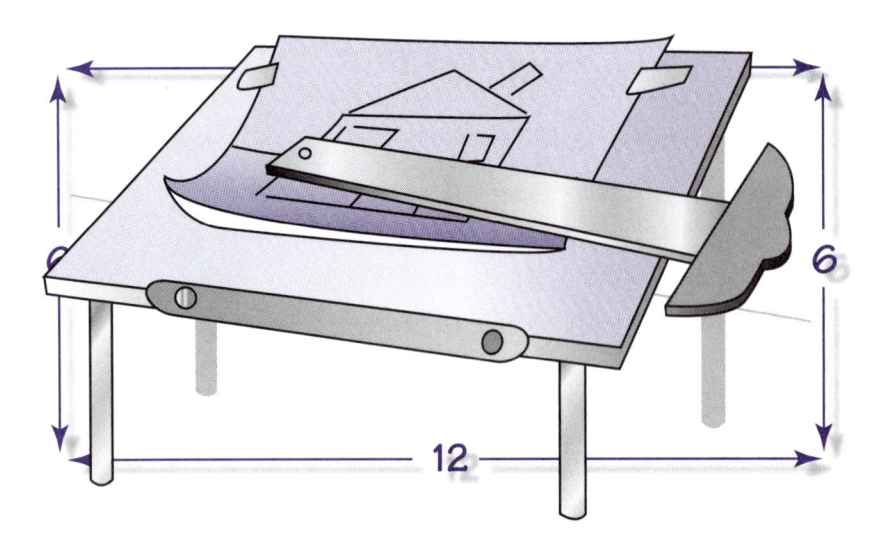

Architects and drivers use scale drawings! Who else does?

Scale: 2 in. = 3 ft

1. 12 ft
2. 1½ ft
3. 6 ft
4. 21 ft
5. 9 ft
6. 7½ ft

Scale: 5 cm = 1 m

7. 10 m
8. 0.2 m
9. 3 m
10. 0.5 m
11. 8 m
12. 1.5 m

Answer Box

A	B	C	D	E	F
50 cm	2.5 cm	15 cm	6 in.	7.5 cm	1 cm
G	H	I	J	K	L
8 in.	14 in.	40 cm	4 in.	1 in.	5 in.

Objective: Determine the scale drawing length given the actual distance.